Wind Power In My Hometown

A School Teachers Perspective and DIY Model

Andrew G. Swapp

DEDICATION

I dedicate this writing to all who have helped a small town high school shop teacher make his dreams come true. From parents and students, to industry professionals. I especially would like to thank Curtis Whittaker who always provided opportunity for me and my students and took a back seat when it came to accolades and publicity. With out Curt I would have nothing of this magnitude to write. Of course my wife and family have been very supportive and have been full of encouragement. I thank them for helping me all along the way.

CONTENTS

ACKNOWLEDGMENTS

I would like to acknowledge the Whittaker family who catapulted me and my students into a wind hall of fame if you will. Without their including us in the prospecting journey of finding commercial scale wind for development. I would have much less to write about.

I must acknowledge First Wind, Paul Gaynor, Krista Kisch, Peter Sullivan, Brian Harris, and Bill Dent and his crew. A huge thanks for the friendly acceptance of student involvement and thoughtful action in our behalf.

To all of Beaver County for the support of a teacher who may look out the window a bit too much. To all of the opposition that made us think and become a little more responsible in doing things that effect many.

To my students who trudge along with me in experiments that bring them out in the heat of summer and the frozen cold of winter. I appreciate all of the extra mile you have gone for this to happen.

1 MY INTRODUCTION TO WIND POWER

The wind has been used for years to do work and has been relied upon for many different things from pumping water to producing electricity. In the age of rural electrification wind power was almost forgotten, with exception of the cattle rancher in the western deserts who depended on the wind to pump water out of the ground to keep cattle alive and thriving. Wind energy has withstood the litmus test of politics and economics and is now the fastest growing form of energy development in the world today. Wind energy has proven over time and with many technological advances to be an economically viable and clean resource. Wind energy is also a tool for teaching Technology Education. Students are concerned about the future of energy, environment, and a comfortable lifestyle. Introducing them to wind energy gives them an avenue to explore that fulfills the need of having a secure future.

I really didn't know any of this upon graduating from my Technology and Engineering Teacher courses at the university. I didn't study renewable energy in college, but I had a keen interest and did study it on my own. I was introduced to passive solar design at an early age when on a family vacation my mother saw a strange looking house. Of course, she pulled right in and parked the car. I crouched down in the back seat and was embarrassed. We had no idea about who lived there and what they were like. I was a very shy kid who was raised in Southern Utah in a small rural town and somewhat sheltered to the world. Mom motioned to me to get out of the car and follow her to the door. I sheepishly did as she wished. She pointed out how thick the walls were and how much glass was in the front. I started to forget about being shy and started getting very interested in this strange house.

A knock on the door brought a long haired man wearing cut-off shorts and sandals. He asked what we needed and Mom said that we were just traveling through town and noticed the design of his house and we just wondered if we could find out more about it. A smile shown through his long beard and he said come on in! The house was so simple yet beautiful. I looked with awe at the full glass front and thick adobe walls. We learned that the heating of the house was done with the sun and a wood cook stove. He told us that it only took one quarter of a cord of wood all winter. I was hooked on this design and wondered why more people didn't do this. With the combination of the sun heating the floor and walls and a little bit of wood a house is almost completely self sufficient. This is significant information. He told us thermal mass is equivalent or better than insulation.

I will never forget the comfortable feeling in that home.

Years later. I found myself guiding passengers on river trips through the Grand Canyon on the Colorado River. I was amazed at the Anasazi Indian dwellings and granaries. They had passive solar and thermal mass down. There houses were earth berm or practically under the ground and the granaries were located on the North face of cliffs in sealed with mud and rock. These were placed so that they would be shaded year-round. The seeds stored would keep for very long time.

As a soldier traveling around other countries, I noticed the lack of a power grid of any kind and the lack of effectiveness if there was one. In tropical climates people had very little need for electric power. I met a family in Honduras in 1988 while on a patrol from our regimental support area. They had a mango tree so I had to stop in and bargain for a few mangoes. As we approached the house, I noticed that the house had no glass in the window and no door in the doorway. The older couple were sitting out on the front porch, or porch area, where the home built adobe stove was. I noticed the U.S. Army T ration trays that obviously were dug from our garbage and were now used as cook ware. The walls of the home were of very thick mud and had a thatched roof. I felt the difference in temperature as I was welcomed to the porch the shade of the roof and coolness of the cement or mud floor and walls were very inviting. We asked for some mangoes and they told us to take all we could eat. They were just going to waste. Then I spotted the largest avocado I had ever seen in my life. The lady saw me looking at it and she shook her finger saying no, no that is not to give away. I pulled out one lempira and was looking for a couple more and she snatched the money from my hand and gave me the avocado. I was embarrassed at the low amount that I given her and tried to give her more and she refused to take it so I left it on the porch wall when I left. When I recall standing next to that shaded wall it was the most comfortable and coolest I had ever been in Honduras. I was so glad that they welcomed a small contingent of U.S. Army soldiers to there home. We were in full combat gear with cammo face paint and weapons at the ready. They must have thought us to be a bit strange or scary.

All of these events keep adding up to the mind set that there must be something in these claims of efficiency in alternative building and renewable energy. This stuff was always on the back of my mind. It never really hit the forefront while I was traveling the world and serving my country as a soldier.

I guess my introduction to wind power was in the 1980's when I got stationed at Ft. Ord California. I would travel home to Utah on long weekends and on leave. The most direct route took me through Tehachapi, California where I was amazed at the wind turbines covering the hill sides. The first time through that area I had to pull off the road and just watch them turn. It was hard to judge the size and scale of the turbines until I saw a service truck parked by one. They were not as big as the turbines of today, but they seemed huge back then. There were some with three blades and some that looked like big egg beaters. I was committed to the U.S. Army at that time and gave them no further thought other than they were cool and it was a neat thing that they were doing.

2 SAND BLASTING MY BARN

While still in college I was taking a break from home work and I picked up the classified add paper called the Pioneer shopper. I found and eighty acre farm with all the equipment and bordering BLM land. This would have been a dream if I had already graduated and if I had a job. I looked on for entertainment and the add kept nagging at me. I turned back to it and read it again and again. I yelled into the kitchen, "Sharla, I'm going to call on this farm."

She chuckled a bit and said, "Go ahead we don't have any money to buy anything like that anyway!"

She felt secure that it wouldn't go any further than a phone call. I called and it sounded good, but I got the feeling that I should ask my friend to go look at it for me. I called Steve and he said, "What are you doing Andy? you can't make money on a small hay farm. The price of hay is about the same as it was when you and I were in high school!"

I said, "That I know, but just go look at it for me."

I hung up and went about my business. The next day I had several messages on my answer machine when I got home from school. "

Andy you need to come look at this place it has two houses all the farm equipment and a huge barn, if you don't buy I will!"

This was shocking news. I thought and thought. I don't have any way to buy it.

I called Steve and he said, "Make an offer contingent on getting financing."

The offer was what I thought to be low for value I did see in the small farm, but it was what I though I might qualify for. I shot the offer and they countered for just a few thousand more and I said okay. My next two weeks were spent in every bank, lending institution, and Farm Service Agency. It seemed impossible to find a loan officer who wasn't a flat world thinker. Finally I found one. She worked out the detail and with help from my friend Steve, we purchased the farm. When our home in St. George sold we paid Steve back and will forever be grateful for his help.

I moved in and started to get familiar with things. I was so anxious to farm that I started up the old David Brown tractor and began to work a twelve acre field. A light snow fell on the freshly turned soil and then in the next few days the winds picked up and the soil picked up in the wind as well. The sandy loam soil traveling at such a speed in the wind literally sand blasted the paint off of my barn. The once white barn was showing silver now. I wanted to get mad, but instead I thought there is real power in this wind. Someone really should put up wind turbines and do something with this stuff! In fact I should see about getting a little turbine and powering some of my farm. I really didn't think any more about it for a while.

I was re-introduced to the idea of wind power when I was driving around doing errands with my father in-law when we heard an add on the radio that the state energy office was doing a free seminar on wind power that evening. We looked at each other and said, "Lets go!"

We found ourselves enthralled with the presentation. I approached the engineer after the presentation and asked if she thought a farm and school project would be considered viable. She seemed hesitant at first and then she agreed that it may be a viable thing. I put in our application for a MET tower or meteorological tower to be erected on my farm. The application was accepted and a date set of October 2001. I was so excited!

3 MY FIRST MET TOWER

The phone call came. The State Engineer called and set a date that she would bring the 20 meter tower to my farm. I didn't understand what exactly 20 meters would look like. I could think in feet and inches so I calculated 39 inches per meter. So at 65 feet I can kind of picture that in my mind. This would be a fairly tall tower. I thought it would be a good project for my students. I just had to decide which students since I taught seventh thru twelfth grades.

My eighth grade class seemed to have a lot on the ball and were in a pre-engineering class. So I arranged for a bus and told the students about what we were going to do. We prepared by getting some tools ready and soon the day came. We loaded the bus and headed down the all to familiar road to my farm. The engineers truck was sitting on the location. The eighteen acre plot was covered in grass and sage brush. The bus pulled off of the road and we off loaded. The engineer had us unload the ten foot tower sections and lay them out on the ground. We slipped one section into the other and propped it up on an old mild crate and a tire. We followed her instruction to attach the instruments. She then called us in to gather around the hood of her truck and she explained how we were to measure out the four anchor points and get them square. She showed us the Pythagorean Theorem. The hypotenuse was the distance from anchor to anchor, where the (a) would be from one anchor into the tower base and (b) would be from the tower base back out to an anchor point.

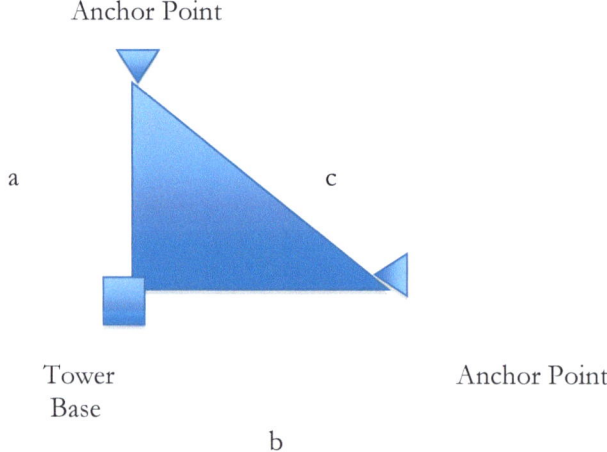

Anchor Point

a   c

Tower   Anchor Point
Base

b

This illustration includes two anchor points of a four anchor point system.

She also showed us how this same formula is used to calculate the length of the guy wires that stretch from the anchors up to different locations on the tower. We did the calculations, measured the distances from the base to the anchors, and began pounding the anchor points into the ground. I had eighth grade students along with my own son who was a sixth grader, swinging sledge hammers and measuring cable out for guy wires. They had a ball!

The moment of truth came when we hooked up the gin pole to the base and to the cable on one side then to the winch on the front of the truck. The button was pushed and the cable started to reel in. The gin pole was standing straight up in the air and it started to come down to the ground or toward the truck. As the gin pole came down it brought the tower up. To see this 20 meter tower come off the old milk crate and snake just a bit as it slowly climbed to an upright position was an awesome sight. This is the time you must trust your calculations and your measurements. If the side cable are too short or too long the tower may lean so much that it could kink and fall over. I must say that this part of the build was just a bit intimidating. I had my students standing off to the side with tag lines keeping the tower in line and straight as it raised up. Once the tower was up, guy wires adjusted, the tower straightened by eye, the batteries installed, and instrumentation tested, our install was complete and tested with parameters in place. We were now testing the wind and direction and averaging it every ten minutes and recording it on a data logger.

We loaded the tools back into the engineers truck and thanked her for the great field trip and the tower. She was impressed with the work and enthusiasm of the students. We said goodbye and loaded on the bus. On the ride back into to town, the students gathered to the back of the bus and the excitement was very prevalent in their voices. I then heard them talking about the math and the Pythagorean Theorem, and one of them said, "Yeah, I never thought we would use that crap!" That comment was a crowning moment of glory for a teacher. They found a use for some things they had learned. This is what a shop teacher is all about. Showing students how the math, theory, and formulas are applied.

Within a week I had three different commercial developers or prospectors stop by my home. A couple of them were a bit upset that I had a tower that was visible from the highway. The prospector who was very friendly came to the house and I was out back working on the siding. I remember him asking what I was doing with the MET tower. Then he asked if he could look at the data. I shared the data and we talked a bit more . He got in touch with one of my administrators and arranged to send us a 50 meter MET tower. Talk about an intimidating tower, WOW!

The fifty meter tower arrived and plans were in place for a good location that would sample the wind speeds and direction. With the help of the district maintenance men and the superintendent we put the tower together and started raising it into the air. The superintendent was running the bar that guided the cable onto the winch and the anchor that held the winch pulled out of the ground and the cable whizzed by the and almost hit the super! The tower was only about five feet off the ground and it crashed back down and rested on the straw bales. We inspected the instrumentation and   the tower. Everything checked out so we reconfigured the winch tie down and connected it to the receiver hitch on the service truck. This time the tower came up steadily. This was a huge accomplishment and motivator. We started on professional quality data collection that was legitimate enough to take to the bank. We also had Campbell Scientific who donated the data logger helping us with the installation. This was truly a combined effort to find good wind if there ever was such a thing.

The data was sent to my computer at the high school every morning at 0800. I had no idea what interest this would create in my students. When a storm would blow through the valley I could expect to see students stopping by the shop early in the morning before school started to check on the wind speeds at the MET Tower. This tower became a tremendous impetus to talk to young minds full of curiosity about meteorology, engineering, electronics, and communications. I was so pleasantly surprised to have such a great tool to introduce and spark the imagination about so many great careers. Who would have seen it coming?

The excitement of things that could be were flooding over not only the students but the town. We went out after school and tested the wind with groups of volunteers and hand held instruments. The protocol that Curt Whittaker sent us was a comparative study of the anemometer at the airport to hand held instruments strategically placed across the valley. This hand held data allowed us to see where the main wind currents flow through the valley and map them for future placement of MET towers. The inclusion of other teachers, parents, and business owners spread the excitement to several who would otherwise be uninformed. The data continued to come in and it looked good, but not quite good enough! This meant a move was coming. Curt was looking for the best wind which not only meant high wind speed but duration and time of day.   The total package would be the selling point not just gusts.

We moved the tower from the South end of the valley to the North and West of town. This tower move was done with students and the district maintenance crew. (Side Note- Some of this original installation crew are now wind technicians who are working full time at the wind farm they helped to discover.) This second move did not last very long either so a third move was in order to test for even more wind, more consistent wind. We got the order to move and again we moved this fifty meter giant to a location about eight miles North of town. The numbers started looking very good. Students were still coming in to see the wind speed on particularly windy days. Before long I got the call that a professional team was headed to Milford to place seven more towers and a company called UPC was going to start commercial scale prospecting.

The 20 meter MET tower
put up on the Swapp farm
in 2001 with the help of
Mr. Swapp's eighth grade class

SWAPP'S FIRST MET TOWER
PHOTO BY: ANDY SWAPP

4 COMMERCIAL SCALE PROSPECTING

Our data collection was good enough to attract the attention of a commercial scale development company. Curt Whittaker had selected the UPC group to take over testing and development of the Milford Wind Farm. Paul Gaynor headed this group as the CEO. Paul took on the project and assigned his top developer Krista Kisch to obtain permits and make all necessary arrangements to get MET towers spread across the valley. Mohit Dua, a meteorologist, paid us a visit along with Bob Ochoa, a master of MET tower construction. I lead them on a tour of the valley and they seemed pleased with the ease of access and flat terrain. This was an exciting time to know that a commercial developer was spending this kind of time and money on our valley and basing it from our data collection efforts. The sense of accomplishment for my students and me was incredible.

The process was slow and tedious with data collecting in several areas now and more towers to come. I was hired to check on the towers monthly and to pull data from those not transmitting properly and email it to Mohit. The new sites required Bureau of Land Management BLM permitting. I had the honor of escorting the BLM person in charge of permitting in the area and I found a person that took her job seriously. She wanted the best for the public that she serves. I was amazed at the checklist that was made to ensure proper use of public lands. In time permitting for more locations was obtained and more towers erected.

I had the honor of being the guide to many who came to our valley and played a part in establishing the due process of environmental, anthropological, civil, and governmental requirements. Curt Whittaker made sure that we, Milford High School, were involved in every step of the process. My students were introduced to so many different jobs and occupations through this process. I was so pleased with the educational value that had come from this event. We were invited to present a renewable energy video at the states first renewable energy conference at the Salt Palace a year prior to all of this. My students presented their video to over five hundred people. They were a captive audience. We were featured at lunchtime! From that meeting our network of friends and supporters grew exponentially. The state energy engineer who helped us set up our first MET tower was responsible for giving us the exposure to most of the renewable energy professionals in the state. We were perceived as students and teacher doing a great thing.

As we were walking to the parking lot at the end of the day. People would come up and ask my students if they were the ones who presented at the luncheon earlier in the day. They were congratulated and thanked for their efforts. When we got in the car one of the students said "Wow this was great I have never felt so much like a celebrity in my life!"

I myself wasn't out to save the planet or green-up Utah. The state that has coal for its state rock. I was after jobs for my students when they graduate. I was after some economic growth being brought to a dying little town, but most of all I wanted a real world project for my kids to be involved in and see the benefit of following their dreams. I have no problem with being a little bit cleaner and renewable energy is clean and it is just a cool technology that stimulates interest in students. I was so glad to see, hear, and feel the success that my students were having at this point in the development. We had a joyful three and half hour ride home from Salt Lake City.

I would mention the great possibility of having commercial scale wind in our valley and I was still laughed at by some, but as the progress grew at a steady pace and my students talked about it around town and to their parents the excitement grew. Eventually the day came for a ground breaking ceremony.

50 METER MET TOWER ON THE WAY UP
PHOTO BY: ANDY SWAPP

5 DEVELOPMENT WORK

Finally, after miles of dirt roads checking towers, briefing government and civic groups, and lots of fence post leaning discussions, the ground breaking time was here. My students and I were invited out to the site where a large tent was set up. Moments before the speeches were given, I was asked if I would get up and say a few words. I hesitantly and excitedly agreed. I wanted to do it for recognition of my students yet for my own personal preference I would rather be in the field and away from the flag pole. The speeches were given and everything seemed to go so smooth, but again there was no mention of Curt Whittaker and all the work that he and his family had put in. I mentioned the support to our class and thanked him for his consideration of the education aspect of the project. I never quite understood his standing in the shadows and just watching the accolades pass him by. I knew that Curt was a worker, an intellect, and an influencer, I grew to have a great respect for the man.

We ventured outside where shinny chrome shovels awaited a large group of dignitaries. A student and I were asked to take spot behind a shiny shovel. I was so honored and the students were astonished, for lack of a better word. We were part of the actual beginning of construction of what seemed like just a dream a few years prior.

The surveys were done, roads marked, tower locations set, and heavy equipment perched at the ready. A cement batch plant was moved in and a fleet of concrete trucks ready to go. The job site soon turned into an ant hill of activity. A lot of local talent was put to work moving dirt, holding survey rods, putting large electrical cable in the ground, and placing rebar reinforcement into specially designed foundations. The excitement meter was pegged. I was in a state of wonder. Now there were a hundred more occupations for us to explore and learn about.

Prior to this, the all naysaying stopped because the first tower pieces rolled through town on specially built transport trucks. Not that there was much naysaying at all at this point in time, but the reality was here. Turbine towers rolled into town one after another. Then blades came on even more specialized transport. The length and size of the blades required the Highway Patrol to stop traffic so that the truck could use the whole road to get over the overpass coming into town. Students would come into class and report to me that several more blades just arrived.

The work of tower erecting is something we never have given much thought to, but we learned that a wind tower is one of most highly engineered structures in the built world. A wind turbine tower has to withstand all of the normal forces, such as, compression, shear, extension, and it has to withstand all of the forces that a moving object on top can exert. The turbine spinning and yawing will create torque and even a gyroscopic force on the tower. I don't really know of any other structure that can host all of those forces at same time. The construction of these towers is so very specialized. The people who build these towers are a special breed and come from different walks of life.

GROUND BREAKING OF THE MILFORD WIND
FARM FIRST PHASE
PHOTO BY: MHS STUDENT

6 PHASE ONE BUILD

The build had started while I was still in session at the school and then the month of June would be spent in at the university working on my master's degree so I would miss out on a lot of building. I thought it would be great if I could get connected with the construction company and learn all about how the towers are built and how things go together. I contacted my professors and they said it would be an awesome opportunity to work on my project so I all I would have to do is register for three credits at the university and do a paper on my experiences and I could pick up my studies the next summer. I was pumped! I drove out to the job site the next morning and asked if I could fill out an application to be a laborer. They had no idea who I was, or how involved I had been in getting the wind farm to this stage. I really didn't think about it much and they treated me the same as anyone applying for a job. I walked out of the job trailer when I was done filling out the application and I saw some white hard hats walking my way to I stood there and Mike Welch, the job site foreman, stopped and asked me if I needed a job. I said yes I am the local shop teacher at the high school and I would work for nothing just to learn how everything goes together. It stopped him for a second and he said, "Shop teacher, well you can crank a wrench can't you?"

I said, "Yes I can crank a wrench!"

"I need to put you on a torque crew."

The next two days I waited to get to work with the torque crew but the headman on the job asked me to take his truck around and get photos of the construction. I enjoyed this a great deal, but still wanted to get my hands dirty. The next thing I knew I was in a crew truck headed out to put slew bearings on rotor blades using torque equipment that I had never seen before.

I was put on Justin's crew. Now this was one top-notch crew who prided themselves on speed and accuracy. They knew how to work hard and keep a since of humor. I worked with this crew for about a week and then some of us were asked to help with rotor build. So we went over to the site where cranes were picking up giant rotor blades and plugging them into the rotor.

I was amazed at the accuracy the crane operators had. They were within a fraction of an inch with their movements. Inside the hub you would see the blade slowly come up and the studs line up to the holes in the hub and bam! The blade was pegged to the hub and we would spin the nuts onto the studs and torque them down in a specified pattern. It didn't take us long to get very proficient at this task and before long we were building three rotors in a day. We were pushing other crews to get done so that we could build the rotor and have it ready to lift in place when the tower was built. I was asked at that time if I wanted some experience on the GE wind turbines so I went over and helped install base and mid towers.

I started working under job foreman Terry Lee. I enjoyed the base mid job we hooked up towers to the crane, tipped them upright and moved them into place and bolted them down. We would tighten them a bit, but a tensioning crew would come in and tension the main tower bolts that secured the tower to the foundation. I did get some experience with DTUs or Down Tower Units and then I was put on top out with Robby Lowe as the foreman.

Robby knew that I wanted the whole gambit of experience so he sent me up tower with the young guys. Usually the top out crew was reserved for the young and very healthy. I seemed to handle the ladder climb just fine and heights didn't really bother me too much. The days got very windy so they switched us to a night crew. The night was quiet and calm, perfect for lifting things way up in the air with the 777 ton crane.

On my last job, before I had to quit and go back to school, I got to bolt the top tower piece on an eighty meter tower, bolt on the nacelle, and rotor blades of a Clipper 2.5MW turbine. The young former marine and lead man was teasing me about having to go out on the top and cut away the crane from its last lift of the night. One of the crewmembers was very empathetic and kept saying, " No don't make teach go out and do that, I'll go out again if I have to."

I then said, "I would like that."

So at 0230 in the morning I attached the monkey strap and went out on top and down to the hub where the crane was attached and unscrewed the giant turn buckles and let the crane free from the last lift of the day. I then went back and got the impact wrench and put the covers on the lift tabs. This work at night, 265 feet above the ground was so peaceful and quiet until I made noise. It seemed like a no-mans land almost, where very few have tread.

I felt it an honor to be so trusted as to be a part of the crews that build America. These men work in the worst of conditions at any hour and any time of day or night. I had worked with some of the best in the business and learned a bunch to take back to my school and to finish up my degree. Most importantly I made a few new friends who will always be among the best of memory.

Andrew G. Swapp

CRANE AWAITING MORE PARTS TO ASSEMBLE
PHOTO BY: ANDY SWAPP

7 PUTTING FORMER STUDENTS TO WORK

With a new wind farm practically in my back yard now It was time to get some of the original wind kids a job. I couldn't go out and apply for them and I couldn't just ask First Wind to hire these guys, but what I could do was write a letter of recommendation for them and give them encouragement. Now I have several of my former students are working at the Milford Wind Farm and in the wind industry.

If there is anything that gauges a teachers success it may well be the success of your students. I have felt very successful in this regard. I have presently just over fifty people who I have taught, mentored, or encouraged to apply and work in the wind industry. I think back on the welders, the carpenters, and the engineers I have trained and there are none of my former students doing any of those things full time as a living yet. I do have some up and coming engineers, but the wind has been very successful as far as providing work for my students.

I know of one such student who was a 4.0 student, an athlete, and a very hard worker. The scholarships did not come rolling in for him, but college was in the game plan. He went to work for the wind turbine construction company for two years then went to school. After making good money with lots of overtime he could afford to go to school and school really means something to him now. When school gets hard he just thinks back to working in fifteen below zero weather and his problems seem to disappear. That growing up time and being disciplined enough to save money while working on the road has benefitted him greatly.

Another student who went into wind farm construction found himself working on cell towers in the off season. He has transferred his skills so that he can keep working even in the times when the wind slows down a bit. The same student has worked in the tool room on a job site as well as every other phase of construction. Talk about a valuable asset to have on a job site now.

I get very excited to hear the stories and feel of the success of these students and parents of students who have followed the construction and actually made a living doing so. I am honored to be asked to write letters of recommendation and to answer phone calls from future employers and talk to them about my students attributes. I can only hope that they live up to my high expectations and ideals.

For those who are not ready for the workforce and are set on more formal education I have suggested scholarships. The renewable energy companies that we have worked with in the past have been more than willing to work with me on this.

CONSTRUCTION WORKERS PUTTING A BEARING ON A BLADE
PHOTO BY: ANDY SWAPP

8 SCHOLARSHIPS AND MORE

Early on in the prospecting I was asked by some, "What do you get out of all this?"

I had to think for a minute and say, "I get the excitement of seeing my students go to work. I get the pleasure of learning more about the process of siting, building, and running a wind farm than most people would ever know. I have rubbed shoulders with some of the finest men anywhere from dirt movers, tower builders, and CEO's of major companies. I have been in the company of senators and congressmen. I have been invited to speak in L.A., Chicago, and Washington DC. Although I would rather not speak, it was a pleasure to recognize my students and their accomplishments."

In retrospect, First Wind did hire me as a consultant at the rate of $500.00 a month, I would check MET towers, escort contractors to the site, and act as general liaison for the project. This little side job was very helpful and it allowed me to meet all the professionals that played a part in the development of the wind farm. Last but not least, I get the pleasure of encouraging students to apply for scholarships that I have urged into existence and helped to design.

The scholarships that have graced our school have been a true motivator to teacher and student. When companies care enough to give an annual scholarship to our school, it shows a willingness to be a good neighbor and friend to the community. I have the responsibility of teaching a renewable energy class at the school and this class does give the students an advantage when applying for the scholarships. Scholarships are a great thing but it is something when big companies move into a small town and boost their economy and create jobs on the periphery.

In our county we have about five percent of our high school graduates go on to college. In a very good year, we have about twelve percent of our graduates go on to college. With this accounting some cringe and say wow why so little interest in college. Others say I never had any college and I have done just fine. We have farmers, ranchers, and railroaders who may have some training or schooling of sorts but a lot of them are high earners with no college degree. The scholarships make access to an education that may otherwise be unobtainable. They do work for the their scholarships in many ways. The work comes in the form of a school career filled with achievement. The students who mature into a work habit and study habit and who play sports and volunteer at the food bank or other form of service, have a long history of accomplishment.

9 THE RESULTS

The results of many people's efforts has been, a 306MW wind farm and a permit to build approximately ten new solar farms in our valley. The main and lasting result of our early efforts is staggering in means of physical structure, new tax base for a small county, and economic growth, not to mention a morale boost to a dying town.

With that being said, the magnitude of impact is difficult to measure with all of the chain reaction or domino effect that has taken place. There has been an increase in our county tax base by over two Million dollars a year. As one fine state senator once said, 'This is the single greatest tax base increase in the county's history!" There was an eighty million dollar increase of money flow in the state during the construction of the Milford Wind Farm. The slow economy wasn't exactly felt during this time in the little town of Milford, Utah.

During the construction of the wind farm, one of our teachers had daughter diagnosed with cancer and the school put on a charity basketball tournament. The RMT construction guys along with many others participated and donated generously to the cause.

Rental units in town filled up and RV parks over flowed. A service station was remodeled and a sandwich shop added. Students were excited and dreaming of which job they would like to do rather than wondering if there were going to be any jobs at all. The Circle Four hog farms leased a lot of land to the wind farm and it benefitted greatly their bottom line.

I personally have almost fifty students and people I have mentored working in construction of or maintaining commercial scale wind turbines. I have a handful working in the commercial scale solar installation and a number of students that have gone into an engineering field in college. I have always measured my success by the success of my students. I have felt tremendously successful the past few years. I am so pleased with the interest, excitement, and determination of my students. They have an enviable approach to life and especially the wind industry. Now my numbers may not seem that large and significant until you consider the fact that some of our graduating classes at Milford High School number around 23.

The town still hasn't grown that much, but that was in the forecast! We do have a new business in town, a construction company has been kept busy for five or six years now, and we have a new hospital at the edge of town. This is a lot of life breathed into existence in a short time for an area that sees approximately one new house built every four years. Just this year alone we are seeing several new homes being built, remodeling, and additions to homes. The positive effects are being felt by many. In my estimation the result has been incredibly positive.

Wind turbines should not block the views of existing homes and views of national parks and scenic byways. If wind developers are smart they will chose wisely where they sight their towers, but for me you can put wind turbines in my back yard any time!

If you have wind in your hometown don't just get angry at it, use it! If you are looking at a small home size turbine of some kind do not go out and buy a bunch of test equipment. A small wind turbine would probably be cheaper and would be re-salable if you did not have the kind of wind to justify the purchase where test equipment would not be easily re-salable.

For commercial scale wind, you will need credible 50meter data to attract a wind farm developer. You want to check with your state energy office and see if they have an anemometer loan program. This would be a viable way to test the wind but often the state equipment comes with an agreement to make the data public. Some prospectors would prefer proprietary data because prospecting for wind is a lot like prospecting for gold. You don't want the word to get out until you can stake your claim.

I hope that you find a pattern to follow in your windy valley or hillside. Remember that you don't need to wait for someone to do it some day. You can do it today! Take charge and make it happen. Good luck!

A three cup anemometer   *A small 900 Watt wind turbine*

ABOUT THE AUTHOR

Andrew G. Swapp was born and raised in a small town in Southern Utah, Kanab. He moved to Emery County in 1979, for his senior year in high school, where he enlisted in the U.S. Army after graduation and had a career until his medical retirement in 1996. With the help of the Veterans Administration a degree was obtained in 2000 and a shop teacher was born. Andy and Sharla bought a small farm and settled in Beaver County where Andy is able to teach school and foster his brand of education. He also has a small hay operation that provides hay for his animals for the winter, and a form of relaxation. He is active in the local LDS church and loves to help others find joy in serving.

For speaking engagements contact Andy: swappsre@gmail.com

www.ingramcontent.com/pod-product-compliance
Lightning Source LLC
Chambersburg PA
CBHW041615180526
45159CB00002BC/863